MOST SECRET

3

Photographic reproduction of the original covers, classified "Most Secret" but otherwise completely unmarked save for the stamped numeral "3", designating it as the third in Gubbins' series of guerrilla warfare pamphlets.
NAA: A3269, Q3

HOW TO USE
HIGH EXPLOSIVES

M.I.(R), THE WAR OFFICE

MAY, 1939

ISBN-13: 978-1976340185
ISBN-10: 1976340187

(©) C.A. Brown, 2017
All Rights Reserved

M.I.(R)

CONTENTS

Introductory vii
How To Use High Explosives 1
Detonators 2
The Time Fuze 2
Laying the Charge 3
Method of Firing 4
Firing Electrically 4
To Prepare the Circuit For Demolition 5
Firing More than one Charge Electrically ... 6
Quantity of Explosives Required 7
Precautions When Laying Charges 11
Destruction of Bridges 11
Destruction of Bridge Piers 15
Destruction of Railways 16
Destruction of Buildings 18
Blocking of Roads 19
Placing Mine Crater Charges by the Camouflet Method 20
Figures 16

M.I.(R)

Introductory.

Military Intelligence (Research), or *M.I.(R)*, was the rather bland designation for what would later evolve into the "dirty tricks" department of the British War Office. Established in Spring of 1939, M.I.(R) was, tasked with development of various special operations, military intelligence, special weapons and equipment, and irregular warfare concepts in the months immediately preceding the outbreak of the Second World War.

With the crisis in Europe rapidly coming to a head, the organisation's work took on far greater urgency. M.I.(R) was expanded and became responsible for not only covert intelligence and irregular warfare, but also the development of novel anti-tank weapons and tactics in an attempt to blunt the sharp edge of Hitler's state-of-the-art armoured forces which, UK War Office planners had correctly determined, could very well ride roughshod across Europe if left unchecked.

Major Colin Gubbins. Gubbins was an artillery officer who had seen service in the Great War in France, and significantly, had been involved in operations against guerrilla forces during the Russian Civil War in 1918/1919 and in Ireland against the IRA from 1919 to 1921. Later, while posted to the Northwestern Frontier in India, he had become familiar with irregular tribal warfare the use of human intelligence assets. He was perfectly placed to author three handbooks for irregular warfare - *The Art of Guerilla Warfare*, *The Partisan Leader's Handbook* and this one - *How to Use High Explosives,* which was co-written with M.I.(R)'s demolition section chief Milis Jefferis.

Gubbins had the three handbooks completed by May of 1939, and then was dispatched to Poland to investigate the establishment of a "stay-behind" guerrilla force in the likely event of a German invasion of that country. As it stood, the lightning speed of the German invasion overtook the planning of the Polish "stay behind" force and Gubbins returned to the UK.

With the outbreak of war, Gubbins was tasked

with establishing the UK's Independent Companies (later designated "Commandos"), which he joined in operations in Norway. Later still, Gubbins was tasked with the establishment of a "Stay-Behind" guerrilla force in the UK in advance of a threatened Nazi invasion of Britain. This force was given the rather nondescript designation of "Auxiliaries" and were nominally placed on the war establishment of the Local Defence Volunteers (Home Guard). The Auxiliaries were formed into local bands called Patrols and were trained in commando tactics and supplied with a variety of weapons and demolitions stores which were cached around the countryside. Operating from underground "operational base" dugouts, patrols would harry and hinder occupying German forces while gathering intelligence which would be transmitted to the British government in exile in Canada via specially trained covert wireless units.

Gubbins was posted to the newly formed Special Operations Executive toward the end of 1940. M.I.(R) in 1940 was merged with Section D of the Secret Intelligence Service to become the Special Operations Executive (SOE). An unusual organisation, SOE combined intelligence functions and special warfare, making it a self-contained covert force which later inspired the creation of the US Office of Strategic Services (OSS), which later itself inspired the creation of the United States Central Intelligence Agency (CIA). In 1943, Gubbins was appointed head of the SOE and oversaw operations all over the world, from Europe to North Africa to South East Asia and the China/Burma/India theatre.

How to Use High Explosives was a short training pamphlet which deals completely with the use of explosives in a guerrilla warfare setting. It was designed as a reference for personnel with little-to-no training in demolitions, providing guidance on the types and amounts of explosives to use to destroy different types of targets, firing devices, electric and non-electring firing trains and circuits. .

The pamphlet was later used as the basis for the SOE's demolitions training syllabus. *How to Use*

M.I.(R)

High Explosives is unique in that it represents the first official British military doctrine on the use of explosives in a guerrilla warfare environment.

CA Brown
September 2017

HOW TO USE HIGH EXPLOSIVES

M.I.(R)

HOW TO USE HIGH EXPLOSIVES.

(1) The following explosives are in general use: -

(a) Ammonal – grey powder – must be kept dry or it will not explode; use two or three detonators tied together if it is damp.

A good explosive for mines buried in the ground, but only half as powerful as (b) or (c) for cutting.

(b) Gelignite and other similar gelatinous explosives. Unaffected by water, good for all demolitions, disintegrates in heat. Keep in a cool place.

(c) Dynamite and plastic explosive – similar to (b) but more like putty in substance. Dynamite disintegrates in water and is dangerous to use in cold weather when it becomes crystalline in substance.

(b) and (c) are made up in sticks, (a) in tins. Blasting powder and gun powder are only of use when buried in a mine and are then only half as powerful as the explosives mentioned above.

(2) Explosives, although very inflammable, are not dangerous to handle. They will not go off by being dropped or shaken.

Small quantities burn quietly in the open, larger quantities may explode due to the heat generated by burning.

Do not smoke near explosives.

Do not use steel instruments when laying charges or a spark may be struck.

Detonators.

Keep detonators separate.

(3) The only way of setting off high explosives is by means of a detonator. The detonator is a small copper tube about 6mm. in diameter and 5cms. long. One end is closed, the other open. It is half filled with a very sensitive explosive. Care must therefore be taken in handling detonators.

The Time Fuze.

(4) The fuze is used to set off the detonator. It consists of a slow burning composition covered with a cotton sheath. One end is inserted in the open end of the detonator and the other end is cut off on a long slant and can be lit with a match. Time fuze burns at the rate of one centimetre a second. The time it will take to burn through can therefore be calculated

from this. In order that the fuze will not pull out of the open end of the detonator too easily, the copper tubing at the extreme end may be pinched into the fuze with a pair of pliers.

Laying the Charge.

(5)   When it is required to lay a charge against any metal, machinery, etc., the high explosive must be packed close against it and tied in position with a piece of string or held there by any other means. Each packet of explosive must be close against the next. In the case of gelignite or dynamite, the end of one packet should be opened and a hole made in the explosive by means of a piece of wood of the correct size to allow the detonator to be inserted in it. In the case of ammonal, the detonator must be buried in the powder. If the ammonal is at all damp, tie two more detonators to the one with the fuze in it.   The detonator, with fuze attached as described above, must be pushed carefully into the stick of explosive prepared for it. The detonator is held in position by pressing the explosive round it with the fingers. As a further precaution it may be tied in with a piece of string. Always remember when cutting steel that

the explosive must be in very close contact with it.

Method of Firing.

(6) The outer end of the fuze is lit with an ordinary match. The match is held firmly against the composition showing where the fuze has been cut. The match is struck by striking the box against it.

Firing Electrically.

(7) A charge of explosive can be set off from a distance at any chosen moment by using electricity.

This is done by means of an "electric fuze head". This appliance is cylindrical in shape and fits into the open end of a detonator. Two wires project from the cylinder, when they are connected to a battery a small quantity of gunpowder is ignited which is sufficient to set off the detonator.

To fire a charge electrically the following stores are required: -

(a) An electric fuze head.

M.I.(R)

(b) Two long insulated electric wires such as are used for house lighting, 100 metres will do.

(c) A motor car battery.

(d) Insulating tape.

To prepare the circuit for demolition.

(8)   Join one long wire on each short wire of the fuze head, by twisting them together, and bind the joints with insulating tape. Place the fuze head in the detonator so that its end comes close to the explosive in it. Make a hole in one of the packets of explosive and press the detonator into it.

When it is required to fire the charge the two ends of the long wires will each be pressed firmly against the two terminals of the battery, one wire to each terminal.

Before joining wires together or to the battery, remove the insulation for 3cms. and scrape the wire clean and bright.

A good torch battery in place of the car battery will fire one detonator.

Firing more than one charge electrically.
(see Fig. 1)

(9) It is often necessary to fire two or more charges at the same moment. For instance when several charges are placed on a steel girder, if one charge went off first it might easily displace the others.

This can be most easily done electrically. It must be understood that electricity flows from one terminal of a battery to the other, only if a continuous metal path is provided for it.

The wires must be so connected that the electricity so flowing will in turn pass through each of the several fuze heads it is required to fire.

To do this, one long wire is first joined to one of the wires in No. 1 "fuze head". The other wire in No. 1 "fuze head" is joined by a short length of wire to one of the wires in No. 2 "fuze head". The other wire in No. 2 "fuze head" is joined by a short length of wire to No. 3 "fuze head" and so on. The second wire in the last "fuze head" is finally joined to the second long wire which leads back to the battery.

Again when the two free ends of the long wires are pressed against the two terminals of the battery all the charges will be fired at once.

M.I.(R)

A car battery with 3 cells will fire two or three detonators and with 6 cells will fire five or six detonators, but the long wires must not be too thin or sufficient electricity will not flow to make the filaments of the "fuze heads" red hot.

If there is any uncertainty and spare "fuze heads" are available, connect up the wires as described above, without putting the "fuze heads" in the detonators, and see if the current is sufficient to burn the filaments. The "fuze heads" will have to be replaced by new ones after the test.

Quantity of Explosives Required.

(10) The quantities given below refer to gelignite, and other blasting explosives. Double the amount of ammonal when using it to cut steel or masonry.

(a) <u>Cutting steel rails.</u>
½ kgms. of explosive placed against the side the rail will cut it. 1½ kgms. are required if the explosive is underneath and touching the rail. Fig. 2.

If the explosive is buried under the rail not more than 10 cms. from it 4 kgms. must be used.

(b) <u>Cutting Steel Plates and Girders.</u>

The rule for the amount of explosive required to cut a steel plate is :-

5 gms. Of explosive for every cm. width of plate 1 cm. thick.

For a plate twice as thick (2 cms.) use four times as much (2 x 2).

For a plate three times as thick (3 cms.) use nine times as much (3 x 3), and so on.

(c) <u>Masonry Walls.</u>

The rule for masonry walls and arches is similar to that for steel :-

10 kgms. of explosive must be used for every metre length of wall, one metre thick. If the wall is twice as thick four times as much explosive must be used (2 x 2 = 4). If it is required to cut an arch by placing a charge on its lower side when the upper side is strengthened by earth filling – the thickness to be cut should be taken as the thickness of the masonry plus half the thickness of the earth.

M.I.(R)

(d) <u>Craters.</u>

20 kgms. of explosive buried 2 metres deep or 40 kgms. of explosive 3 metres deep will do. The width of the crater will be about three times the depth the explosive is buried.

(e) Concrete reinforced with steel is very difficult to cut with one charge. It is best to shatter the concrete by placing a charge equal to double that required for masonry and then to cut the steel bars with separate charges placed close against them. The rule for cutting steel bars or steel wire ropes is:-

50 gms. of explosive will cut a bar 1 cm. thick. For a bar 2 cms. thick use four times as much.

$$(2 \times 2 = 4).$$

For a bar of 3 cms. thick use nine times as much.

$$(3 \times 3 = 9).$$

Example of calculation of explosive required to cut steel.

(11) (a) It is required to cut a steel plate 80 cms. wide and 7 cms. thick.

We know that :-

50 gms. are required for a plate 10 cms. wide and 1 cm. thick.

For a plate 80 cms. wide and 1 cm. thick.
50 x 8 = 400 gms. are needed.
For a plate 60 cms. wide and 7 cms. thick.
400 gms. x 7 x 7 = 20 kgms. are needed.

Example of cutting a masonry arch loaded with earth by a charge places on its lower face.

(b) A masonry arch bridge is 10 metres wide and the thickness of the masonry at the place it is to be cut is 60 cms. There is a thickness of earth of 2.20 metres over the arch.

Thickness to be taken for calculation is :-

60 ÷ 2.20 x ¼ = 1.70 metres.

For 1 metre width and 1 metre thickness 10 kgms. are required.

For 10 metre width and 1 metre thickness 100 kgms. are required.

For 10 metre width and 1.70 metre thickness 100 x 1.70 x 1.7 kgms. = 290 kgms. are required.

If the earth were dug away from the inside and the explosives placed on top the thickness to be cut would only be .60 metres.

For 1 metre with and 1 metre thickness use 10 kgms.

For 10 metre width and 1 metre thickness use 100 kgms.

For 10 metre width and .60 metre thickness use 6 x 6 x 100 = 36 kgms.

Precautions when laying charges.

(12)   (a) Always carry detonators in their box and keep them separate from other explosives.

(b) Do not smoke near explosives.

(c) Do not use a steel pick or shovel near explosives or a spark caused by striking a rock may set off the charge.

(d) The last operation when laying a charge is always to put the detonator into the explosive.

(e) When connecting up a charge which is to be fired electrically, always make quite certain that the battery is nowhere near the ends of the wires before everyone has got to a safe distance.

(f) Before firing a charge a plan must always be made so that everyone in the party is at a safe distance.

THE DESTRUCTION OF BRIDGES.

(13)   The easiest way to destroy a bridge is to bury a large crater charge behind the wall carrying the girder or arch as the case may be. Read the paragraph on craters again. The hole must be dug so that the distance to the wall or arch to be blown down is not more than the depth that the explosive is buried. The charge

measured in kgms. should be twice that used for an ordinary crater.

With a wide bridge it may be necessary to put in two charges and fire both together electrically.

Steel girders may be cut by explosives but this is the most difficult type of demolition to tackle and should not be attempted except by men who have had experience.

A steel girder consists of a top boom, cross bracing or plates in the centre and a bottom boom. The bottom booms must be cut near the centre of the bridge, but this may not bring the bridge down if the roadway and cross bracing hold it together.

The only way to make certain is to cut both booms and the cross bracing of both girders. To do this six charges and sometimes more must be fired at once. If the charges are fired separately the first charge to go off will displace the other charges. Read the paragraph on firing more than one charge electrically.

The best way to place the explosive against the steel-work of a bridge is to tie the explosive on to a piece of wood and to tie the wood to the girder with the explosive against the steel.

In the case of a steel girder which has a projection at the top and bottom, it will be best to tie a piece of wood against these projections so as to leave a cavity into which the explosives can be filled. If this cavity is too big it must be made smaller by filling some of it with clay, bricks or blocks of wood. See Fig. 3.

If the drawings are studied carefully it will be seen how this is best done.

Example of explosive placed against a steel girder.

(14) A diagram showing how to fix a charge against a steel girder may be found on the last pages – Fig. 3.

The girder is 20 cms. wide and 120 cms. deep. The thickness of the top and bottom flanges is 4 cms. The centre plate is 2 cms. thick. The calculations of the amount of explosive required are as follows :-

Explosives required to cut top and bottom steel.

5 gms. cuts 1 cm. wide an 1 cm. thick.
5 x 20 gms. cuts 20 cms. wide and 1 cm. thick.

4 x 20 x 4 x 4 cuts 20 cms. wide and 4 cms. thick.

(15) Because the explosive is going to be put all on one side and not spread equally over the steel, use twice this amount, 5 kgms. at the top and 3½ kgms. at the bottom.

Explosive required to cut centre plate.
 50 gms. cuts 1 cm. wide and 1 cm. thick.
 50 x 12 gms. cuts 120 cms. wide and 1 cm. thick.
 50 x 12 x 2 x 2 cuts 120 cms. wide and 2 cms. thick
 = 2½ kgms.
  Total explosive 9½ kgms.

(16) The best way of demolishing a masonry arch bridge is to dig down through the roadway until the masonry is reached at a point close to the abutment right across the road.

The charge is calculated from the rule given in paragraph 9 (c). This amount of explosive is spread over the width of the arch, detonators and fuze or detonators and "fuze heads" for electric firing inserted and the trench refilled with earth.

The Destruction of Bridge Piers.

(17) A bridge pier can be cut by a charge of high explosive placed against one side. The calculations for the amount of explosive required can be found in Paragraph 9 (c). Half as much again should be used, however, on account of the weight carried by the pier and the class of workmanship used in this form of construction, making it more difficult to demolish.

It will be found that a very large quantity of explosive is required to demolish a pier by this means. For instance 1,350 kgms. would be required to cut a pier 10 metres long and 3 metres thick.

Many piers are, however, constructed with a chamber in the centre for demolition purposes. If the explosive is put in this chamber it will be seen that the thickness of the wall will be less than half and the amount of explosive required less than a quarter.

The best method of placing a charge against a wall or pier is to tie a large piece of timber against it so as to form a shelf. The explosive can then be laid out on this shelf, spread equally along the whole length.

DESTRUCTION OF RAILWAYS.

(18)   In attacking railways the best results can be obtained by cutting the rails so as to de-rail a train. This is best done where the railway passes over a bridge or is on a high bank or the railway is cut into the side of a hill.

The simple de-railment of a locomotive will not, in the ordinary way, cause much damage unless a large quantity of explosive is used as in method 4 below, or the rail is cut at some point as described above.

The following are 4 methods of de-railing a train. See also Fig. 2.

1. ½ kgm. of high explosive placed against the side of each rail.
2. 1½ kgm. of high explosive placed against the underside of each rail.
3. 5 kgms. of high explosive buried under the ballast not more than 10 cms. from each rail.
4. A single charge of high explosive buried 1 metre deep between the rails. This will lift the locomotive high into the air and is the best way where no bridge or steep slope can be found.

If the de-railing is done by methods (3) or (4), or where the ballast has been allowed to come close up under the rails by method (2) as well, it will be undetected by day. Care must be taken not to show any signs of digging. A tin of water should be carried to wash down the stone ballast and clean it of earth adhering to it when using methods (3) or (4).

In all cases it is best to fire the charge under or just in front of the front wheels of the locomotive.

This can be done in two ways :-

(a) By means of an electric detonator with long wires leading to a battery where a man is concealed to operate it at the right moment.

(b) By means of a striker machine which is buried under a sleeper next to a rail joint. The weight of a locomotive passing over releases a striker which fires the charge by means of an instantaneous fuze.

In both cases the detonator must be buried firmly in the explosive. When a battery is used, great care must be taken that the battery does not come near the ends of the wire till the last moment to avoid accidents.

A length of wire up to 100 metres may be used leading away from the explosive to a hidden spot where it is fired.

Insulated wire, such as is used for electric light in houses, must be used. The accumulator battery out of a car is best, but a good hand torch dry battery will do.

Diagrams of the methods (1) and (2) of cutting rails are to be found on the last pages – Fig. 2.

The Destruction of Buildings.

(19) To demolish a building, holes should be dug about 1 metre in depth and charges placed in all the corners inside the walls. If there are strong cross walls, charges should also be placed where these meet the outside walls. The amount of explosive required for an ordinary building is 1 kgm. for every square metre of floor area. If it is not possible to dig holes for the explosive, twice this amount should be used. If the walls are more than 50 cms. thick, the charge must be increased accordingly.

Blocking of Roads.

(20) It is not easy to use high explosives for blocking roads as in the case of railways. The difference being that diversions can be made much more easily in the case of a road. It is therefore necessary to choose a place where it is difficult to make a diversion. Such places will usually be found where the road is carried over a bridge or marsh or along a steeply sloped mountain side. In the last case, the best site to choose is where heavy walling has had to be used to hold the road up, and where the ground is soft.

Blocking of the road can best be carried out by means of a mine crater – read again Paragraph 10 (d) which gives the amount of explosive required. The explosive should be buried so that it will blow away the retaining wall holding up the road, or else so that it will start a landslide, which will descend upon it. In both cases it will be seen that there are two open faces, one on the side and one at the top. For this reason the explosive is able to form a crater more easily and only two-thirds of the amount need to be used. At the same time, in order to ensure that the earth is blown out sideways and not upwards, the distance from the charge to the vertical face should not exceed the depth at which the charge is buried.

Placing Mine Crater Charges by the Camouflet Method.

(21) The advantage of this method is the increased speed at which crater charges can be placed. The stores required are as follows :-

(a) A length of pipe, equal to the depth at which it is required to place the charge, whose inside diameter is about 5 cms.
(b) A shoe, either of cast iron or hard wood, on the lower end of the pipe.
(c) Heavy hammer with which to drive the pipe into the ground.

Alternatively, a special equipment may be obtained which will do the whole job.

The method of operation is as follows :- The end of the pipe is plugged up by means of the shoe and the pipe is driven into the ground. If the pipe is to be removed, the shoe must slide easily into the bottom end of the pipe and have a projection all round on which the pipe bears. A small charge of about ¼ kgm. is made up with a detonator and fuze and is dropped down the pipe and exploded at the bottom. This bursts through the bottom end of the pipe forming a round cavity into which the pipe leads. The main crater charge can now be dropped into the cavity via the pipe. Half the charge is put in first, followed by the detonator and means of firing it, and then the remaining half of the charge. The mine can then be fired in the ordinary way.

If the special equipment is used, the pipe is extracted 3ft. after it has first been driven into the ground. The cast iron shoe is so made that the pipe comes away with it leaving it at the bottom of the hole. The small camouflet charge is now dropped in and fired. By this means it will be found that the pipe is undamaged, and can be removed completely after the main charge has been filled in. The great advantage of this method is that once the pipe is removed, there is no trace of where the charge has been placed.

Figures.

ELECTRIC DETONATORS — Figure. 1.

DETONATOR

FUZE HEAD ELECTRIC

INSULATING WIRES

METHOD OF FIRING MORE THAN ONE CHARGE

DETONATORS FITTED WITH FUZE HEADS

SHORT BRIDGE WIRE

ALL FOUR JOINTS CLEANED KNOTTED TOGETHER AND COVERED WITH INSULATING MATERIAL

TWO LONG WIRES TO BATTERY

THE BARE ENDS OF WIRE AND BATTERY TERMINALS MUST BE CLEANED AND CONTACT MADE AT THE APPROPRIATE MOMENT.

METHOD 1

Figure 2.

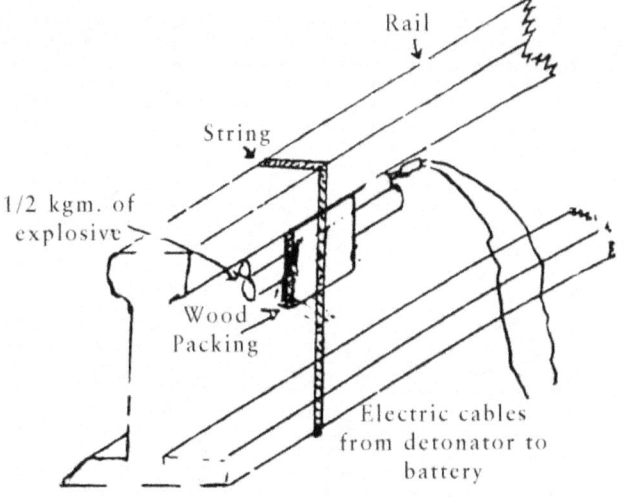

METHOD II

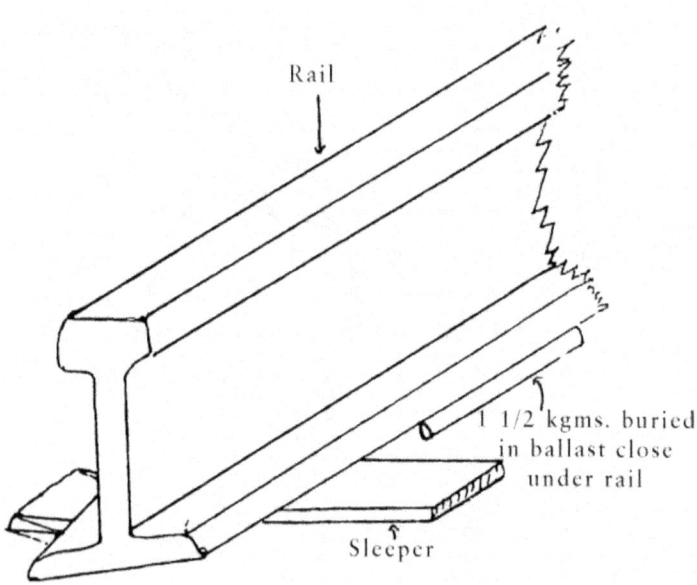

M.I.(R)

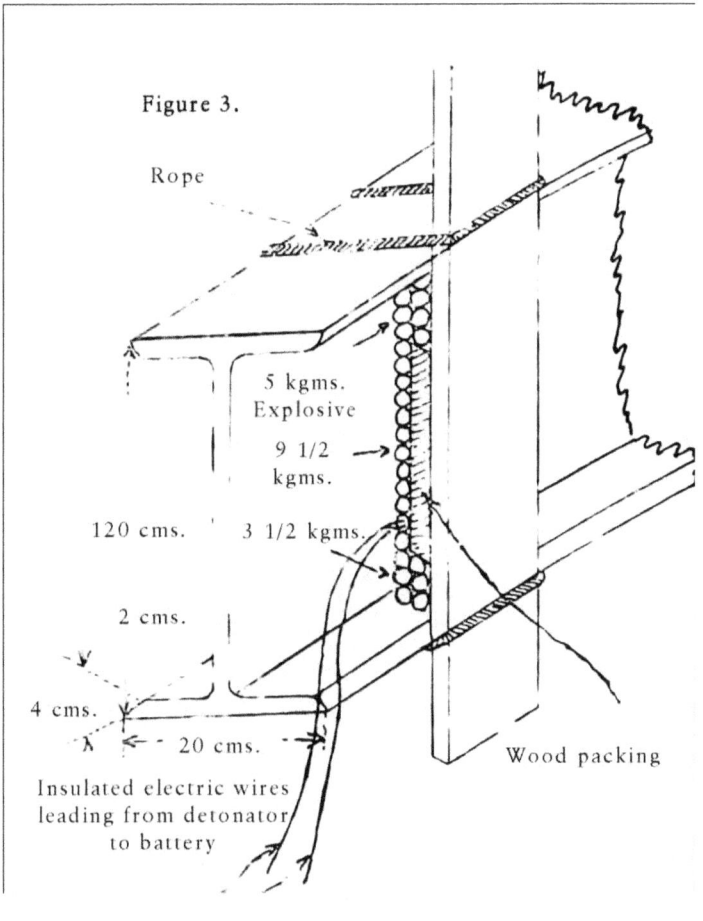

Figure 3.

www.ingramcontent.com/pod-product-compliance
Lightning Source LLC
Chambersburg PA
CBHW050251230526
45470CB00005B/2218